Janeth Mosquera
Abel Muñoz

Design of Different Intraspecific Mixtures in Maize Genotypes

Janeth Mosquera
Abel Muñoz

Design of Different Intraspecific Mixtures in Maize Genotypes

Reducing damage from foliar pests and diseases

ScienciaScripts

Imprint
Any brand names and product names mentioned in this book are subject to trademark, brand or patent protection and are trademarks or registered trademarks of their respective holders. The use of brand names, product names, common names, trade names, product descriptions etc. even without a particular marking in this work is in no way to be construed to mean that such names may be regarded as unrestricted in respect of trademark and brand protection legislation and could thus be used by anyone.

Cover image: www.ingimage.com

This book is a translation from the original published under ISBN 978-613-9-40871-9.

Publisher:
Sciencia Scripts
is a trademark of
Dodo Books Indian Ocean Ltd. and OmniScriptum S.R.L publishing group

120 High Road, East Finchley, London, N2 9ED, United Kingdom
Str. Armeneasca 28/1, office 1, Chisinau MD-2012, Republic of Moldova, Europe
Managing Directors: Ieva Konstantinova, Victoria Ursu
info@omniscriptum.com

Printed at: see last page
ISBN: 978-620-8-51926-1

TOPIC:

DESIGN OF DIFFERENT INTRASPECIFIC MIXTURES IN MAIZE GENOTYPES TO REDUCE DAMAGE FROM FOLIAR PESTS AND DISEASES.

PUYO - ECUADOR 2024

SUMMARY

Maize is one of the crops most affected by a variety of foliar diseases and pests, which is why it is necessary to know control methods. Currently, new agroecological methods have been proposed to protect the plant from diseases and pests, one of which is the use of genetic aspects. Hence, through this research work, the objective of the study was evaluate the effect intraspecific mixtures of genotypes in maize in order to reduce the attack of pests and foliar diseases of the plant. The following variables were evaluated in this research: plant height, ear insertion height, 100-seed weight, ear length, kernel weight per ear, grain to ear ratio and yield. A randomised complete block design applied to ten treatments with three replications was used and Tukey's test was used to compare means. As a result, a positive and effective reaction of the genotype mixtures was determined, due to the reduction in the level of attack of foliar diseases and the level of damage and severity of the budworm. The results obtained allow us to conclude that the use of intraspecific mixtures is efficient and that they should be implemented in maize cultivation to improve productivity yields and generate profits for producers.

Key words: Agroecological management, severity, yield.

I. INTRODUCTION

The cultivation of maize is of great economic importance at a global level, as it is used as a source of food for people, but also for animals or for the production of industrial products. World maize production between 2017 and 2018 was more than 2034.45 tonnes with a yield of 6.75 tonnes per hectare (CIMMYT, 2019).In many countries, such as Brazil, Argentina, the United States, Ecuador, etc., maize is a crop that has been cultivated in many countries. As well as the companies dedicated to offering seeds of this crop, and the farmers who grow maize try to contribute to the conservation and generation of genetic diversity. On the one hand, they maintain traditional local varieties and on the other, they intentionally select the most favourable seeds for their diverse characteristics, through the variants that have been presented the following characteristics: natural selection, mutation, recombination, introduction and isolation and thus come to form new races (Casafe, 2017).

Research has shown that criollo seeds mixed with commercial seeds have very special characteristics such as resistance to drought, frost and diseases. Thus, the use of this type of varieties has become a good alternative increase genetic diversity due to the fact that in these mixtures there is a lot of plasticity and a great adaptation to different environments (Quintana, 2019).

Due to all the above information, the effect of intraspecific mixtures in maize genotypes for the reduction of pests and foliar diseases attack was evaluated in this research work in order to determine the incidence of foliar diseases (Curvularia lunata, Spiroplasma kunkellii, Puccinia sorghi and Helminthosporium spp) in different intraspecific mixtures of maize genotypes, Spiroplasma kunkellii, Puccinia sorghi and Helminthosporium spp) in different maize intraspecific mixtures, as well as the severity of damage by the budworm (Spodoptera frugiperda) in the different treatments applied.

1.1. Problem

The problem of pests such as the budworm (Spodoptera frugiperda) and foliar diseases (lunata, kunkellii, sorghi and Helminthosporium spp.), considerably affect maize cultivation; the use of technological packages comprising chemical fertilisers and pesticides has been gradually introduced in all regions where maize production takes place, thus transforming traditional agricultural production systems into systems dependent on chemical inputs, mainly pesticides. There is scientific evidence that their use seriously damages the environment and public health, not only weakening the quality and fertility of the soil and soil production, but also causing long-term adverse effects on the health of the people who carry out the work and management of this crop. It is necessary to identify agroecological solutions to reduce the use of pesticides by means of techniques or methods that improve productivity together with the corresponding phytosanitary care and management.

1.2. Hypothesis

Intraspecific mixtures of maize will reduce the use of pesticides and thus prevent the attack of foliar pests and diseases.

1.3. Objectives

I.3.1. Overall objective

To evaluate the effect of intraspecific mixtures of genotypes in maize for the reduction of pest and foliar disease attack.

1.3.2. Specific objectives

- To determine the incidence of foliar diseases (Curvularia lunata, Spiroplasma kunkellii, Puccinia sorghi and Helminthosporium spp.) in different intraspecific mixtures of maize.

- To determine the severity of damage by codling moth (Spodoptera frugiperda) in different intraspecific mixtures of maize.
- To propose integrated pest and disease management alternatives for maize through the use of cultivar mixtures with different levels of resistance.

II. THEORETICAL FRAMEWORK

2.1. Conceptual foundation

2.1.1. Agroecology

Agroecology is "a new agriculture that is based on agroecological principles on the intermediate systems of agricultural production by means of components with high biodiversity among plants and organisms fulfilling an ecological role and thus reducing external inputs (Altieri and Nicholls, 2014).

2.1.2. Crops

The practice of sowing seeds in the ground and carrying out the tillage and management of the field in order to obtain seed production (Duke, 2016).

Cultivation is any action carried out by a person with the aim of improving, treating and transforming the land so that the sowings through the planted seeds grow (Yánez et al., 2018).

2.1.3. Plots

A parcel of land refers to a piece of land that is part of a larger tract of land, can be used in different ways and for different purposes (Barg and Armand, 2017).

2.1.4. Edaphoclimatic characteristics

Edaphoclimatic characteristics are related to the soil and climate in order to determine the degree of suitability for planting. It takes into account variables such as altitude, soil textures, geographical location, physical, chemical and biological characteristics of the soil, etc., which are used to determine the areas destined for conservation and protection (Polania et al., 2017).

2.1.5. Sowing

It comprises the action and effect of sowing, i.e., throwing and scattering seeds on prepared land. It also refers to the time of sowing and the land that is sown (González, 2016). It is an agricultural activity involves placing a seed in a soil prepared for that purpose, and in the long term new plants germinate and emerge (Gómez and Minelli, 2017).

2.1.6. Agronomic performance

It is the phenotypic response to a stimulus, which can be rapid and generally also has the possibility of being reversible (Caicedo, 2017). It is the number of expressions that plants show in response to different environmental stimuli and among individuals of the same species (Arregui and Puricelli, 2018).

2.1.7. Fertilisation

It is the process by which various substances, whether natural or chemical, are added to the plant in order to make it more fertile, useful and productive (Naranjo, 2017).

2.1.8. Weed control

"It prevents plant species from growing and stealing nutrients from the plant, and this is achieved through care such as weeding and the application of substances that eradicate them completely" (Duke, 2016). They allow "through specific methods such as weeding, to prevent the appearance or infestation crops by avoiding the use of chemicals as much as possible and to avoid causing harm to the environment and people" (Guerrero, et al., 2017).

2.1.9. Phytosanitary control

This control avoids, prevents or reduces the economic losses that can be generated by the pests present in the pests being grown, by using the most appropriate and suitable measures at all times while the crop is growing (PAN International, 2016).

2.1.10.Pest and disease control

It is "known Integrated Pest and Disease Management, it seeks through different control methods, an approach that seeks to bring together crop-specific conditions in which it is desired to totally eliminate pests and diseases from the cultivated plants" (Devine et al., 2018).

It allows crops to be kept in good shape, so that damage from pests and diseases is below the economically acceptable level. It reduces the risk to human health, the environment and the cost of production for those who grow crops (Alavanja, 2019).

2.1.11.Genotypes

A genotype in terms of agriculture is a collection of genes in a plant. It can in turn refer to the two inherited alleles of a particular gene (CIMMYT, 2019).

It is the genetic information that the plant possesses in the form of DNA, which includes numerous variations of its genes (Ortega, 2017).

2.1.12.Intraspecific mixtures

"Intraspecific mixing comprises the biological interaction between two or more individuals of the same species whose purpose is to interact and achieve benefits between them" (Poehlman and Sleper, 2015). "are carried out by two individuals who are related to each other because they belong to the same species and are established for purposes breeding, feeding and protection against pests and diseases". (Louette and Smale, 2016).

2.1.13.Pests

Any animal, plant or micro-organism that negatively affects agricultural production (Arregui and Puricelli, 2018).

2.1.14.Spodoptera frugiperda Smith (codling moth)

It is a worm that feeds on the bud of the plants, it is the main agent that causes severe damage to the leaf tissue of plants in the growth stage, reducing the leaf area and negatively affecting the photosynthetic rate (Arévalo and Mejía, 2018).

2.1.15.Foliar diseases

They are "diseases that form many lesions on the leaves and stems of plants, causing massive tissue collapse and, consequently, nutritional imbalance of the plant" (González, 2016). "Diseases that act by reducing the green leaf area, as well as the activity and duration of the plant, resulting in decreasing the growth rate of the plant or limiting the availability of leaves during the maturation process of the plant" (Tielemans, et al., 2017).

2.1.16.Treatments

treatments are a specific set of conditions to be imposed on the experimental units of an experiment (Yánez, et al., 2018). comprises the set of actions that are applied on the selected experimental units and that are compared with each other (Bernardino et al., 2019).

2.1.17.Tukey's test

"allows testing of all differences found in the treatments used to test the hypotheses, the number of replicates is required to be constant across all treatments" (Kundu, et al., 2017). It is a method that aims to compare the individual means from an analysis of variance of different samples that have been subjected to different treatments (Aguirre et al., 2019).

2.1.18.Variance

It is a measure of dispersion of a random variable, which is widely used in the area of statistics in which its value is expressed by a number, i.e. the variability of that calculated dispersion (Kundu et al., 2017).

2.1.19.Design of complete blocks.

This compares three sources of variability: the treatment factor, the block factor and random error. Overall, it refers to the fact that in each block there is randomisation and all treatments are tested (Kundu, et al., 2017). "It is used in research where there is a source of variability or where there is a factor that is positively or negatively affecting the application of treatments. This requires that the material is homogeneous and that the distribution is completely random and that there is an equal number of experimental units in each block" (Yánez et al., 2018).

2.1.20.Experimental error

It is a deviation of the measured value of a physical quantity from the true value of that quantity (Yánez et al., 2018).

2.1.21.Davis scale

"It is a very useful tool for decision making, as it can be used to understand the effects or damage caused by a pest on a specific crop or plant" (Goodman and Rawling, 2018).

2.2.Theoretical basis

2.2.1. Maize

Maize is grown in all regions of the world where soil conditions are suitable for agricultural activities and all months of the year. It also grows from 58° north latitude in Canada and Russia to 40° south latitude in the southern hemisphere (Yánez et al., 2018). It is cultivated in regions below sea level and at altitudes of over 4,000 metres in the Peruvian Andes. But despite the great diversity of its forms, it seems that all the main types of maize known today, classified as Zea mays, were already being cultivated by native populations when the Americas were discovered. Whatever the case, most modern maize varieties are derived from material obtained in the southern United States, Mexico, Central and South America (Yánez et al., 2018).

2.2.2. Taxonomy of maize

The taxonomy of maize is as follows: (CIMMYT, 2019).

Kingdom: Plant

Division: Tracheophita

Subdivision: Pterapsidae

Class: Angiosperm

Subclass: Monocotyledone

Order: Gumiflorales-Graminales

Family: Poaceae

Subfamily: Panicoideae

Tribe: Maidaea

Genus: Zea

Species: Zea mays

Common name: Corn

2.2.3. Botanical description of maize

According to Louette and Smale (2016) maize has the following botanical description:

- **Roots:** These are fasciculated and provide a perfect anchorage for the plant. In some cases, a few knots protrude from the roots at ground level, and this usually occurs in secondary or adventitious roots.
- **Stem: It** is simple, straight, cane-shaped and solid inside, with a tall length of up to 4 metres, robust and without branches.
- **Leaves:** Long, lanceolate, alternate, parallelinervate and large. They embrace the stem and are hairy on the upper side.

• **Inflorescence:** There are separate male and female inflorescences on the same plant. The male inflorescence is a yellow panicle, while the female inflorescence, when fertilised by pollen grains and known as the cob, has the seeds grouped along an axis, and is covered by layers of green leaves.

• **Kernel:** The seed coat (fruit) is called pericarp, it is hard, underneath is the aleurone layer that gives colour to the kernel, it has proteins and in its interior is the endosperm with 85-90% of the weight of the kernel. The embryo is formed by the radicle and the plumule.

2.2.4. Phenological stages of maize

Maize is divided into two stages: Vegetative stage and reproductive stage, shown in Figure 1 (Caicedo, 2017).

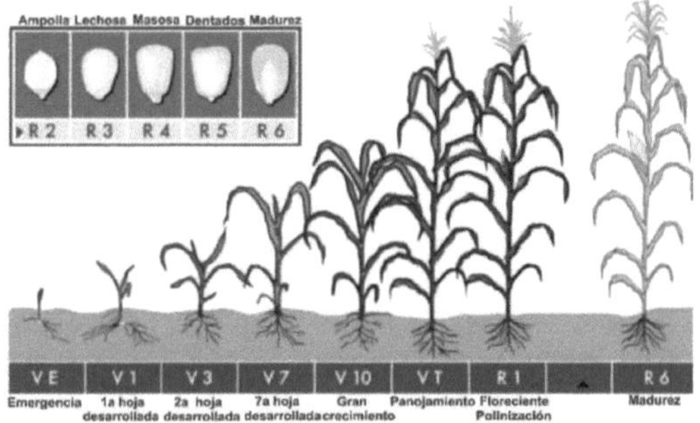

Figure 1. Phenological stages of the vegetative and reproductive phase of maize. **Source:** (Caicedo, 2017). Maize crop management.

2.2.5. Maize phenological vegetative stage

This phase starts at sowing and lasts until shortly before the reproductive structures appear, i.e. when the ear begins to show. of maize (male flower). During the seedling stage any damage to foliage or roots is critical and puts seedling survival at risk (Aguirre et al., 2019).

In the vegetative phase most of the energy is directed to the formation of foliage; therefore, the plant has a certain tolerance to loss of foliage due to pest attack.

According to Goodman and Rawling (2018), this state has several sub-stages, as shown below:

- **Stage VE (germination and emergence):** arrives when the coleoptile sprouts from the surface. Maize plants can emerge within 5-7 days after planting when temperature and moisture conditions are ideal. And under cold and wet conditions or even under very dry conditions they may take more than two weeks to emerge.
- **Stage VC - First leaf:** A leaf with visible ligule (structure at the base of the lamina). The tip of the first leaf in maize is rounded. From this time until flowering (R1), the vegetative stages are defined by the leaf with visible ligule located at the top of the plant. The growing point is below the surface until last part of the V5 stage (five leaves).
- **Stage V2 - Second leaf -** Nodal roots begin to emerge below ground. Seminal roots begin to senesce. The likelihood of frost damage to seedlings is low, except for extreme cold conditions.
- **Stage V3 (three true leaves):** In V3, the growing point is still below the surface. The stem has not elongated much. Root hairs are growing from the nodal roots as the seminal roots stop growing. All leaf and spike shoots which the plant will produce are formed from V3 to V5. A small spike is formed at the end of the point of decrease. The height of the aerial part of the plant is about 20 cm.
- **Stage V6 (six true leaves):** The growing point and spike rise above the soil surface near stage V6. The stem begins to elongate. Nodal root system grows from the lowest 3 or 4 nodes of the stem. Some spikelet shoots or tillers are visible. The development of tillers (tillers) depends on the variety, stocking density, fertility and other conditions.
- **Stage V9 (nine true leaves)**: Dissection of a plant at this stage shows several ear buds (potential ears). These develop at all nodes of the aerial part, except the last 6 to 8 nodes below the ear. The lower ear shoots grow fast at first, but only one or two of the higher ones develop a harvestable ear. The ear begins to develop rapidly. The stems lengthen as the internodes grow. For V10, the time between new leaf stages is shortened to about two to three days.

Around V10, a rapid increase in nutrients and dry matter accumulation begins. This continues throughout the reproductive stages. Soil water and nutrient requirements are very high. This is to meet increased demand due to the high growth rate.

• **Stage V10 (ten true leaves):** the plant begins a rapid increase in dry matter accumulation that will continue until the advanced reproductive stage. High amounts of nutrients and soil water are required to meet the demand.

• **Stage V12 (twelve true leaves):** Although the potential ears are formed just before panicle formation (V5), the number of rows in each ear and ear size are established at V12. However, the determination of the number of ovules (potential grains) will not be completed until one week before the emergence of beards or close to V17.

• **Stage VT (BREEDING)**: This stage starts approximately 2-3 days before the emergence of the beard, during which time the maize plant has reached its final height and pollen release begins. The time between VT and R1 can vary considerably depending on cultivar and environmental conditions.

During this vegetative growth, pests such as thrips (Frankiniella williamsi), larvae of diabrotica (Diabrotica virgifera zeae), codling moth (Spodoptera frugiperda), armyworm (Spodoptera exigua), cutworm (Agrotis sp.), wireworm (Agriotes sp.), blind hen (Phyllophaga sp., Cyclocephala sp., Diplotaxis sp., Anomala sp. and Macrodactylus sp.), weevil (Geraeus senilis), leafhopper (Sphenarium sp., Diplotaxis sp., Anomala sp., and Macrodactylus sp, Cyclocephala sp., Diplotaxis sp., Anomala sp. and Macrodactylus sp.), weevil (Geraeus senilis), leafhopper (Sphenarium purpurascens) and spider mite (Oligonychus mexicanus and Tetranychus sp.) as shown in Figure 2 (Aguirre et al., 2019).

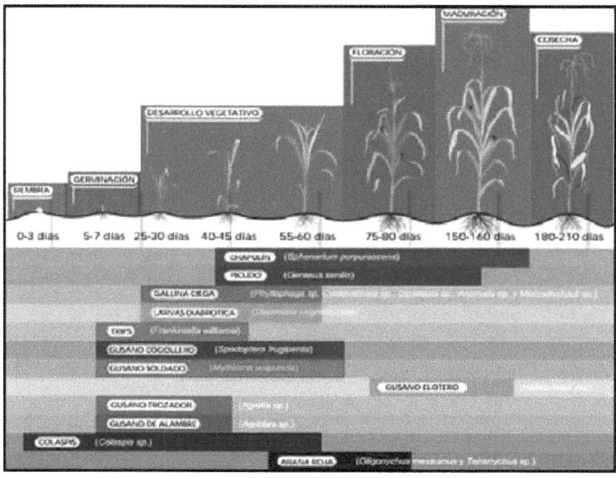

Figure 2. Incidence of main pests in relation to the phenological development of the maize crop.

2.2.6. Reproductive phenological stage of maize

The phenological stage of maize begins when the ear of maize is visible and ends when the crop reaches physiological maturity (black layer at the point where the grain is inserted with the corn). During this stage, pests such as the weevil, leafhopper, red spider mite and corn earworm (Helicoverpa zea) are present (Bernardino et al., 2019).The incidence of pests during vegetative growth is reflected in the reproductive phase of maize, causing large losses in yield potential, due to a reduction in the supply of photosynthates for grain growth. Corn earworm attack on maize kernels can lead to infection by the fungus Aspergillus flavus, which is responsible for the production of aflatoxins, substances that are highly carcinogenic to humans (Bernardino et al., 2019).

2.2.7. Soil and climatic requirements of maize

The maize plant can thrive in regions with rainfall modules of 450-900 mm, well distributed over the growing season. However, even this amount is not sufficient if moisture cannot be stored in the soil, either due shallow soil depth or runoff, or if evaporation is high due to high temperatures and low relative humidity (Brush, 2018).

Table 1. Edaphoclimatic requirements for maize cultivation

Adapt.	Temp. °C	Precp. mm/cycle	Text.	Depth (cm)	(%)	PH
Optimo	19-24	700-850	Franco	+60	-15	Neutral
Good	15-19 24-28	500-700	850-1000	Franco sandy	40-60 15-30	Acid
Marginal	+28	-500 y +1000	Franc or clay so	----------	30 or more	Acid/Alc alginate

Source: (Brush, 2018).

2.2.8. Maize physiology

As more leaves are exposed to sunlight, the rate of dry matter accumulation gradually increases, under favourable conditions the aerial parts of the plants will continue at a constant daily rate until maturity is reached. Highest yields are obtained when environmental conditions are favourable at all growth stages. Unfavourable conditions in the early stages will limit leaf size (photosynthetic engine) (Brush, 2018).At later stages, unfavourable conditions affect the normal growth of the corn, reducing the number of styles and resulting poor pollination of the ovules, which reduces the number of kernels per ear (Brush, 2018). The total growth period of the plant can be divided into two periods: from emergence to the appearance of the beards and from here to physiological maturity, with the first period being the one that can be most affected by factors such as temperature and humidity. The plant's demand for water and nutrients is directly related to the increase in dry matter and decreases in the stages following grain formation (Brush, 2018). The knowledge of the different periods of growth and development of the plant is important for an adequate agronomic management, which are divided into the following periods:

- Germination and seedling establishment.
- Vegetative development.
- Differentiation of panicle and ear.
- Flowering.
- Development and maturation of the grain (Caicedo, 2017),

2.2.9. Cultural work

2.2.9.1. Soil preparation

Soil preparation in maize cultivation depends on the production system in each region. The main objective of tillage is to favour those natural processes that create the most favourable conditions in the soil for seed germination and plant growth (Maffei and Bueno, 2018). The harrow pass depends on the type of soil. In loam-textured soils, the most advisable is one pass and in clay soils, two or three passes can be made (given the characteristics of their aggregates). From a conservation point of view, soils must be ploughed to a minimum, without losing the objective of the crop (Maffei and Bueno, 2018).

2.2.9.2. Sowing

The sowing season for maize cultivation depends on the rainfall in the area or region. For the Pacific coast, the sowing seasons are known as the first and last, as these are the most favourable ecological conditions for production. Especially in terms of the lower incidence of reduction, a disease that is widespread in the Pacific. Two parameters are carried out during sowing:

- **Seed selection:** The use of seeds of creole varieties brings as a consequence in most cases low yields, therefore, it is necessary to promote the use of improved varieties, in order to achieve high yields which requires knowledge of their agronomic characteristics for their adequate recommendation. This means taking the region as the main criterion, since there are different varieties of improved maize in the country that show diverse behaviour, especially in relation to the vegetative period.
- **Stocking density:** Stocking density is conditioned by available moisture in the area or region, natural or induced soil fertility, variety, as well as production use (Maffei and Bueno, 2018).

2.2.9.3. Weed control

Field experiences have shown that losses caused by weeds are of equal or greater magnitude than those caused by insects and diseases; therefore, weed control must be systematic and integrated: cultural control, chemical control, mechanical control (Maffei and Bueno, 2018).

2.2.9.4. Fertilisation

It comprises, according to Maffei and Bueno (2018), two phases, as shown below:

- **Base fertilisation:** At planting time, fertilise with a complete formula, e.g. 12-30-10 at 2 qq/Mz; or 16-36-00-2S, at 2 qq/Mz; complete formulas without K can be used because Pacific soils are rich in this nutrient.

- **Supplementary fertilisation:** The N necessary for the plant and for grain filling must be provided; this can be done in a single application of 3 qq/Mz of Urea at 46% N after 30 days. Although the ideal would be to apply the Urea fractioned, 150 Lbs/Mz at 20 days and 150 Lbs/Mz at 40 days.

2.2.10. Intraspecific biodiversity

From the genetic point of view of the plant kingdom, intraspecific biodiversity refers to diversity within the same species or diversity of genes within a species. For example, barraganete and orito bananas belong to the genus Musa (Caballero et al., 2019). Intraspecific biodiversity is the variation in genes and genotypes between and within species. It is considered to be the sum of the information The genetic diversity contained in the genes of plants, animals and micro-organisms that inhabit the Earth. Diversity within a species allows it to adapt to changes in the environment, climate, agricultural methods used, or to pests and diseases that may affect it (Gómez and Minelli, 2017).Rendón and Aragón (2017) refer to it as the heritable variation within and between populations of a given species or group of species. The genetic diversity that species have allows them to respond and adapt or not to the characteristics or changes in their environment. This is done at the chromosomal level, where recombinations or mutations are made little by little, which can give better or worse adaptive characteristics.Intraspecific biodiversity contributes to the ability of ecological communities to resist or recover from environmental disturbances or changes, including relatively long climatic shifts. Genetic variation in species is the fundamental basis of evolution, the adaptation of wild populations to local environmental conditions, the development of animal species and cultivated species varieties have produced significant direct benefits (Rendón and Aragón, 2017).

2.2.11. Importance of intraspecific biodiversity

It is important because it generates an integral analysis of human beings, since they are part of biodiversity and human society has transformed, transforms and will transform nature. The changes that humans make to nature have to do with individual and collective attitudes (Rimache, 2018). Agroecology is now incorporating intraspecific biodiversity, its management and conservation into the explanation of the structure and function of ecosystems, so that humans can incorporate and return to this new idea of their belonging to nature (Rimache, 2018).The importance of intraspecific biodiversity lies in the past, the present and the future. Because that is where we started some 4 billion years ago. years; in the present, because it is what we focus on today when it comes to obtaining a quality crop and production; and in the future, because it is the commitment of each and every human being to know, protect and conserve diversity when it comes to obtaining a good production (Rimache, 2018).

2.2.12.Loss of intraspecific biodiversity

The causes of biodiversity loss are attributed to local and global market failures. These causes have been caused by a maldistribution of the economic benefit of biotechnology research. , the only beneficiaries of intraspecific biodiversity are those who invested in the research, because small producers or people who own the biodiversity are unaware of the potential hidden in the biological wealth (Navarrete, 2017).As long as they know how to make use of biodiversity, it will be conserved, but as ancestral knowledge is lost and no use is found for the main properties of each plant, genetic erosion will be generated. The fact of possessing the technology means an advantage over those who use biodiversity, even more so if the necessary technology for seed improvement, such as agrobiodiversity, is not available (Navarrete, 2017). Reports now indicate that the Common Agricultural Policy (or CAP) has not been effective in halting the observed decline in intraspecific biodiversity loss. According to them, the number and variety of plant species on agricultural land has been declining for decades (Poehlman and Sleper, 2015). The decline is particularly noticeable in the number of intraspecific varieties in existence today, which is a good indicator loss, having declined by more than 30 % from 1990 values. The loss of agricultural plant diversity is more difficult to assess, although intensive monocultures are a major loss to agricultural biodiversity (Poehlman and Sleper, 2015).

2.2.13.Intraspecific competition

Intraspecific competition generally leads to a reduction in the growth and development of individuals due to changes in the quantities of resources in reserve

(Sánchez and Ruiz, 2016). The density of individuals in space modifies the availability of the light resource, and as individuals move closer together the interactions become more negative.According to intraspecific competition (Salazar and Aldana, 2018) they mention that it is when competitors are of the same species. In addition to that, the downstream effect is a reduction in the contribution of individuals to the next generation, the resource being competed for is limiting, the effects are reciprocal and dense-dependent.

2.2.14.Intraspecific Maize Mixtures

Genetic variation in maize is directly associated with ecological niches where specific environmental conditions prevail. In traditional farming systems, particularly under rainfed conditions, the main genetic input is provided by adapted criollo populations or populations with a broad genetic base (Ortega, 2017).
Traditionally, the conservation of these materials is done through ex situ conservation strategies, however, the management of populations by farmers has been recognised as an important strategy to conserve and exploit their genetic variation (Ortega, 2017).

The ideotype is the programming or a project on the most favourable characteristics that a plant or variety should contain, according to the ecological conditions of a region and its agronomic management (season, number of irrigations, fertilisation, herbicides) and in general what cultivation practices will be used so that the genotype is optimal and thus the phenotype of the variety that is to be formed with the breeding methodology is manifested, which will then be used according to the projected ideotype (Ortega, 2017). The diversity of a species is constituted by all the genetic variations, the product of the difference between species, which can more or less diverse (Gómez and Minelli, 2017).

Genetic variability applies to the characteristics of a population and if there is no genetic variation for that characteristic, the trait cannot be modified by selection. If a change in the environment or living conditions affects that trait, it may disappear (Gomez and Minelli, 2017).

The same author notes that genetic variability is caused by mutations, which are produced by a change in a nucleotide in the sector of the DNA chain that codes for a gene. Generally, individuals of a species differ from each other in many characteristics, and these differences have genetic and environmental causes (Gómez and Minelli, 2017). This is why it is important to mention genetic variance (Vg), because it is a component of genetic diversity. Differences in allele frequencies of a gene governing a trait can create considerable phenotypic variability, such as in morphological traits like fruit colour and shape (Gomez and Minelli, 2017).

Ecuador is one of the countries with the greatest genetic diversity of maize per unit area, and preserving it will represent the most important renewable natural resource for the survival, rural sustainability and food security of future generations.

Up to 2018, 760 entries or collections have been registered in the germplasm bank of the National Department of Phylogenetic Resources and Biotechnology (DENAREF) of INIAP, corresponding to materials of the breeds considered to be high altitude (>2 200 masl) (INIAP, 2008). These collections are classified into 29 varieties, of which 18 are cultivated in the Sierra, characterised by being floury and semi-hard and are distributed according to the preferences of farmers and consumers (Caicedo, 2017). Genetic diversity is not possible to estimate a species in statistical or quantitative terms, because it classifies the species into intraspecific categories such as races or ecotypes; the relative genetic diversity of a species in a region is given in terms of the number of intraspecific categories. Generally, the intraspecific categories of race, ecotype, morphotype and variety are used to classify the diversity of cross-pollinated, wild, agamic and self-pollinated species, respectively. For this reason, each of these terminologies is defined below (Caicedo, 2017).

2.2.15.Race

A breed is an aggregate of populations of a species that have in common morphological and physiological characters and specific uses. However, their distinguishing characteristics are not sufficiently different to constitute a distinct subspecies (Rendón and Aragón, 2017). In the plant kingdom, race classification should be applied only to cultivated species. Breeds are closely related to cultures. For example, maize breeds are part of the cultural heritage of peoples, as are their customs, music, language and many other cultural manifestations (Rendón and Aragón, 2017). Although maize is an allogamous species and therefore there is a great deal of cross-pollination between breeds, which produces many inter-racial hybrids, the breeds can be individualised and universally identified. Everyone can recognise, with a minimum of training and experience, the Tuxpeño breed from Mexico, the Olotón from Guatemala, the Montaña from Colombia, the Chillos from Ecuador, Cusco from Peru, the Kcello from Bolivia, the Cristalino from Chile, the Calchaqui from Argentina, the Avatí Morotí from Paraguay, etc. (Rendón and Aragón, 2017).

2.2.16.Ecotype

It is the product of the adaptation of a species to a particular environment. Ecotype is not synonymous with breed. What defines the ecotype is primarily its area of adaptation. Both breeds and ecotypes are inter-fertile. Ecotypes are occasionally isolated by geographical barriers and are then referred to as geo-ecotypes (Rendón and Aragón, 2017). The term ecotype should be used only for wild species. Scientists collecting wild populations, mainly from forests, use the term "provenance" to indicate the origin of the collected sample. A provenance is not necessarily an ecotype; several different provenances, even very distant from each other, may correspond to the same ecotype (Rendón and Aragón, 2017).

To distinguish between ecotypes, it is necessary to sow all provenances together at one location or at several locations within the adaptation range of a species. Several provenances are grouped within the same ecotype if they show similar morphological characters and physiological reactions (Rendón and Aragón, 2017).

2.2.17.Variety

The term variety to describe the diversity of self-pollinated cultivated species will be used, even though it is known that since 1961, when the Code of Nomenclature for Cultivated Plants was published, the term "cultivar" has been adopted to replace "variety", because the latter is, according to code, very imprecise. The name variety is reserved in the code for certain intraspecific categories of wild natural populations. However, the division of the entire diversity of a species into cultivars does not make sense; most likely, all cultivars of a cultivated species come from a very limited sector of the diversity (Rendón and Aragón, 2017).

2.2.18.Morphotype

In agamic or vegetatively propagated plants, morphotype used to differentiate populations and individuals. A morphotype is defined by a number of characteristics, mainly morphological. A morphotype consists of plants that are morphologically similar; they show the same phenotype, but are not necessarily of the same genetic constitution (Rendón and Aragón, 2017).

2.2.19.Native varieties

Varieties harvested in regions where the crop originated or diversified are called native or autochthonous or traditional varieties, i.e. those varieties that are traditionally used by farmers and have not undergone any systematic and scientifically controlled breeding process, and whose seed is produced by the farmers themselves (Rendón and Aragón, 2017).

Native varieties whose seed is collected and maintained in germplasm banks, duly identified with their origin and geographical location (passport) information, are called "accessions" (Rendón and Aragón, 2017). The objective of genetic improvement is to introduce genetic diversity, the main activity of which is to select the individuals within a population that offer the best characteristics, achieving an accelerated evolution of the selected population. This requires not only phenotypic but also genetic variation (Caicedo, 2017).

The breeding of cultivated plants has one main objective, which is the creation of varieties with high production per unit area, in a certain environment and with certain cultural procedures. Thus, breeding of cultivated species is aimed at breeding materials:

- Resistant to pests, diseases and stem blight.
- Precocious and productive.
- Easy to adapt.
- And nutritionally rich in protein, starch, oils, etc. (Caicedo, 2017).

2.2.20.Multilines and mixtures for disease control

Multilinnias and mixtures are originally intended for the application of protection strategy, although plant protection is nowadays possible by means of other non-genetic control systems. Genetic control uses the genetic variability of plants to obtain varieties with varying degrees of resistance or tolerance to attack by specific pest-producing organisms (Quintana, 2019).

In this sense, genetic selection has been directed towards the development of resistant varieties, resistant rootstocks, the study of certain cultivation techniques such as multilines and mixtures, and the development of transgenic plants (Quintana, 2019). Multilinear varieties, i.e. mixtures of genetically related lines carrying different disease resistance genes, offer farmers an alternative to tame the emergence of crop pests (Quintana, 2019).

Multilines are a combined mechanical-biological option to obtain generalised resistance. Although a new race capable of attacking one of the component lines may emerge, it will not be able to multiply rapidly because that line is surrounded by plants carrying different resistance genes (Quintana, 2019).

A multi-line variety is formed by mechanically mixing seed from several lines that are similar in appearance and genetic constitution, but possess different pest and disease resistance genes. The lines are bred and selected so that they are nearly identical in height, maturity, plant type, grain quality and other characteristics (Quintana, 2019).

On the other hand, Navarrete (2017) describes it as:

- They are mixtures of isogenic lines of self-pollinated species.
- Its primary utility was for disease resistance
- Each line differs in qualitative resistance to the same pathogen.
- The aim is to reduce the likelihood of the pathogen breaking the different alleles for resistance.

2.2.21. Diseases in maize

2.2.21.1. Maize rust (Puccinia sorghi)

This disease, known to attack endemically in the core zone of the maize plant only, causes severe attacks, pustules are formed which cause necrosis of the leaf tissue, giving an appearance of leaf spot, and whole leaves may die if affected. The rust is most noticeable towards flowering. The aetiological agent is a heroic and macrocyclic fungus that apparently cycles on Oxalis spp. (pycnidic and ecidic stages), and the other part (uredosoric and eleutosoric) on maize (Allen, 2017).

Figure 3. Symptoms of the disease Puccinia sorghi

For rust treatment, the use of disease resistant/tolerant hybrids is recommended, as well as the use of natural or organic fungicides (Allen, 2017).

2.2.21.2. Spikenard (Sphacelotheca reiliana)

It is caused by a basidiomycete fungus called Sporisorium reilianum f. sp. zeae (Kühn) Langdon and Fullerton (Synonym of Sphacelotheca reiliana (Kühn) Clinton). It is a soil-surviving pathogen that can infect maize. Reproduction is by teliospores, which are spores with two nuclei at the beginning, shortly afterwards coalesce during the formation of the same. Growth and development only occurs under field conditions or when the growing medium contains maize plant extract. This fungus can remain dormant in the soil for years (Allen, 2017).

Figure 4. Damage caused by Sphacelotheca reiliana

2.2.21.3. Anthracnose stem rot (Colletotrichum graminicola and Glomerella graminicola)

It is a fungal pathogen that attacks the stem of the plant; stress after flowering allows the disease to appear. The way to identify it is the bright or intense black colouring on the stem, it crushes easily when its base is pressed and falls easily if it is pushed. The best way to combat this disease is crop rotation, tillage and using genetic resistance (Allen, 2016).

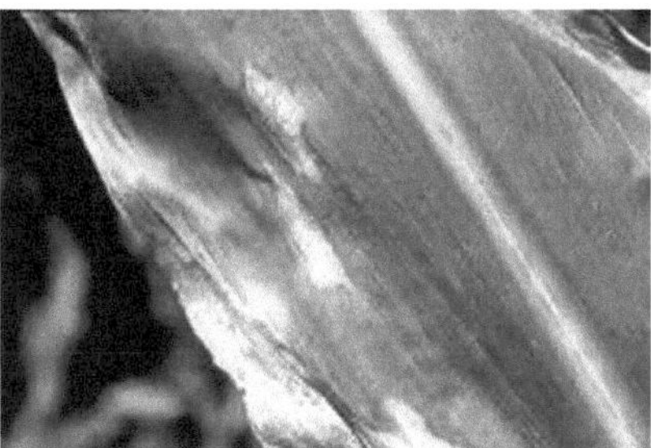

Figure 4. Damage caused by Colletotrichum graminicolae

2.2.21.4. Stem and root rot (Fusarium graminearum, Gibberella zeae, Scierotium bataticola, Macrophomifla phaseoli, Diplodia maydis)

The predominant pathogen is Fusarium graminearum and Gibberella zeae, which cause stem and root rots as the plant approaches maturity. This disease progresses when the plant is stressed by drought, foliar diseases, insect damage, low fertilisation or compaction (Allen, 2016).

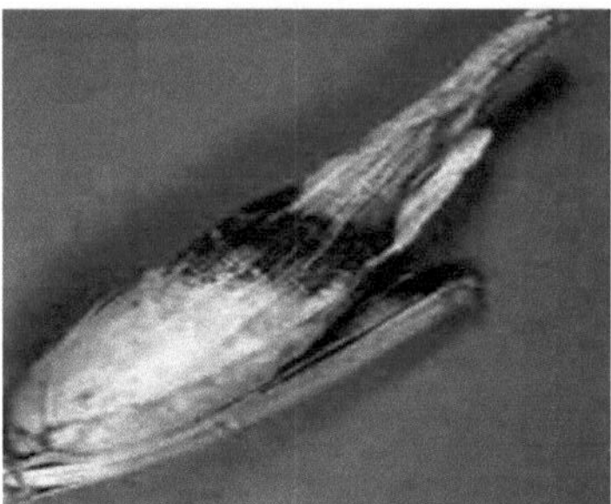

Figure 5. Symptoms of Gibberella zeae disease

2.2.22. Pests in maize cultivation

2.2.22.1. Grey worm (Segetum, Ipsilon, Exclamationis)

Grey worms are larvae of various butterflies that are part of the Noctuidae. They are between 4 and 5 cm in size, rolling up they feel the contact of a possible predator. They are greyish in colour, and in case of Agrotis ipsilon they have black stripes on their rings (RAP-AL, 2017).

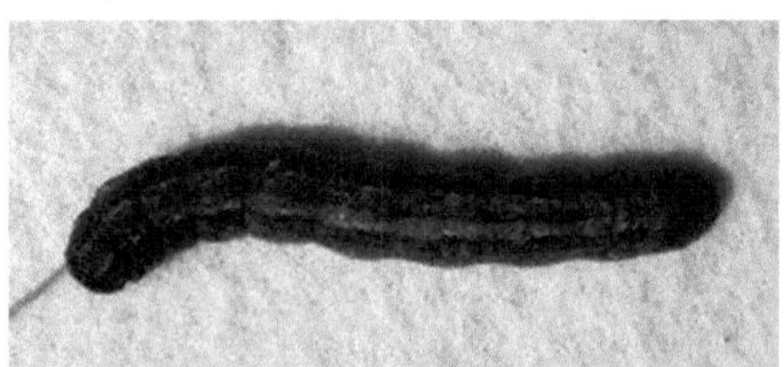

Figure 6. Agrotis ipsilon worm.

The grey grubworm causes damage to the maize crop with larval bites. They cause a generalised wilting of the central leaves on the young plant, spreading over time to the rest of the plant. A heavy attack considerably reduces the volume of plants in a plantation (RAP-AL, 2017). Control of grey grub consists of the application of insecticides (10% w/v (100 g/l) Lambda cyhalothrin, etc.) which in the latter case consists of a foliar spray at concentrations of 0.01-0.02 % (RAP-AL, 2017).

2.2.22.2. Corn borer (Sesamia nonagrioides)

This Sesamia caterpillar feeds on both the ear and the stalk of maize, eating the inside of the stalk that supports the tuft (male flowers), causing it to fall off, and thus stopping fertilisation (RAP-AL, 2017). Maize production drops suddenly. In general, there is some resistance to the borer in adult plants, with only a reduction in yield and quality. In the case of young plants, a severe attack can completely damage the crop (RAP-AL, 2017).

Figure 7. Adult Sesamia nonagrioides causing damage to maize. Treatment against corn borer consists of insecticide application (10% w/v deltamethrin at a dose of .125 L/ha, 48% chlorpyrifos in normal spray, 15-20 cc/10 L water) (RAP-AL, 2017).

2.2.22.3. Seed fly (Phorbia platura)

This dipteran insect is about 0.4-0.6 cm long and is attracted to moist, cool or tilled areas. The larvae of the seed fly develop in soil cavities. This causes problems for the seed. They appear empty or with galleries dug by the larvae. In the event of germination, the plant appears deformed or not very vigorous, because the roots are affected by the larvae (RAP-AL, 2017).

Figure 8. Phorbia platura fly.

Insecticide application (10% w/v (100 g/l) Lambda cyhalothrin, etc.) with an application of 0.01-0.02 % of product is necessary for the control of the seed fly in maize (RAP-AL, 2017).

2.2.22.4. Screwworm (Elasmopalpus angustellus)

The adult corn earworm has a wingspan of 21 to 25 mm. The wings of the adult are greyish in the female and light-coloured in the male. Larvae can be up to 20 mm long and are dark grey with black shades on the head (RAP-AL, 2017).

Figure 9. Damage caused by Elasmopalpus angustellus.

The damage caused by the borer in the maize crop is based on stalk borings. Uniform perforations can be observed on the leaves (RAP-AL, 2017).

2.2.22.5. Maize caterpillar (Heliothis armigera)

Maize caterpillar damage is caused by larval bites on stalks and fruit. Severe attack on the crop results complete mowing of the maize. The caterpillar lays its eggs singly on the underside of the leaf (RAP-AL, 2017).

Figure 10. Adult butterfly Elasmopalpus angustellus.

The solution to the maize caterpillar is the application of insecticides (10% w/v deltamethrin at 0.075-0.125 L/ha, 48% chlorpyrifos at a concentration of 0.15-0.2% (150-200 cc/100 L of water) (RAP-AL, 2017).

2.2.22.6. Maize aphid (Rhopalosiphum maidis)

The maize aphid affects the crop by sucking the plant material, in particular leaves and ears. These attacks cause chlorosis, necrosis and loss of vigour of the plant. If the attack is severe, it often leads to a reduction in the number of kernels on the ear. The time when the aphid attacks maize intensively is from spring to early summer (RAP-AL, 2017).

Figure 11. Rhopalosiphum maidis causing damage to maize.

One treatment against maize aphid is the application of insecticides (10% w/v deltamethrin, at concentrations of 0.075-0.125 L/ha) (RAP-AL, 2017).

2.2.23. Budworm (Spodoptera frugiperda)

This pest is highly relevant for the maize crop, being the most important of all pests, although it has also been detected in some other crops such as beans, onion, alfalfa, tomato, cucumber, among others (Romero, 2019).Among the problems encountered in controlling the pest are the suitable conditions provided to the moth so that it can reproduce and spread easily by having large-scale, massive farms. On the other hand, the indiscriminate use of agrochemicals has made this pest resistant and the effectiveness of the products is low (Romero, 2019). Like the rest of the noctuids, adult activity is nocturnal, so during the night they are attracted to light, especially ultraviolet light. Similar to light attraction, codling moths are attracted to honeydew, which is useful for field monitoring (Romero, 2019).

2.2.23.1. Taxonomy

According to Villamil et al (2017), the taxonomy of the codling moth is:

- **Kingdom:** Animalia
- **Phylum:** Arthropoda
- **Class:** Insecta
- **Order:** Lepidoptera
- **Suborder:** Glossata
- **Infraorder:** Heteroneura
- **(no rank):** Ditrysia
- **Family:** Noctuidae
- **Subfamily:** Xyleninae
- **Genus:** Spodoptera
- **Species:** S. frugiperda

2.2.23.2. Description

The budworm feeds on crops such as corn, cotton, small grains, soybeans, peppers and tomatoes. Damage occurs when the larvae bore into the flowers and fruit of the host plant and feed inside the plant; larvae may also feed on the leaves of host plants. This invasive pest can be found in both field and greenhouse environments (Yanggen et al., 2019).

The budworm has four life stages: egg, larva, pupa and adult. Eggs are very small, with ribs extending longitudinally across their surface (Maya, 2016).

They change from yellowish white to dark brown just before hatching. Larvae can be up to 1.7 inches long and vary in colour from bluish green to chestnut red, with darkening after each moult. Pupae are dark tan to brown and 0.6 to 0.9 inches long (Maya, 2016).

Adults have a wingspan of 1.4 to 1.6 inches and vary in colour. Generally, males are yellowish brown, yellow or light brown, and females are orange-brown (Maya, 2016).

Adults emerge from late March to June and lay eggs on a variety of host plants. Larvae go through five to seven developmental stages. Once mature, larvae drop to the ground and pupate to overwinter in the soil, emerging as moths in the spring (Maya, 2016).

During its life cycle it passes through the egg, larva, pupa and adult stages, and its life cycle is completed in approximately 30 days in summer, 60 days in spring and 90 days in winter (PAN International, 2016).

This species may have overlapping generations, which means that different life stages may be present at the same time. The number of generations per year can vary greatly, depending on the climate. Therefore In general, this pest can have 2-5 generations per year in temperate regions, and up to 11 generations per year in tropical regions (Pesticide Action Network, 2018). Depending on temperatures, the complete pest cycle can last between 30 and 70 days, being shorter in warmer conditions and vice versa. In each generation, the pest cycle is divided into four stages. The duration of these stages varies: (i) as a pupa (barely buried in the soil or on stubble), it lasts between 6-13 days; (ii) as an adult, 6-20 days; (iii) as an egg, between 2-5 days and, (iv) as a larva, between 17-32 days (in this stage it goes through 6-9 instars) (Pesticide Action Network, 2018).

In general, a very high percentage of pupae die when exposed to temperatures below 8°C for short periods, e.g. 15 days. In some northern regions of the country, where conditions are favourable (temperature), the insect will overwinter as a pupa buried in the soil. Towards the centre and south of Argentina's agricultural area, the first infestations would come from migrations from warmer areas (Pesticide Action Network, 2018).

2.2.23.3. Reproduction

A female can oviposit more than 1000 eggs during her reproductive period. These hatch in three to five days: the larvae feed on a small leaf area when they hatch, but in the following days they spread to neighbouring plants, settling in the bud. They have cannibalistic habits, so from the third period onwards only one larva per bud is observed (Ramírez and Lacasaña, 2018).

They go through six developmental stages in 14 to 21 days, on the temperature. The pupal stage occurs in the soil and around 9 to 13 days later, the adult emerges (Corra, 2019). Eggs of the codling moth (Spodoptera frugiperda) take 3 to 5 days to hatch, and are laid in masses of up to 300 eggs in any given area. leaf surface covered by a fine cloth formed from the body scales of the adult female (Corra, 2019).

2.2.23.4. Severity level

The budworm can attack maize from germination to crop maturity. Early attacks can affect vegetative stages of development while late attacks can damage ears (Pirkle, et al., 2016).

The most important damage occurs from the first vegetative stages, although in more advanced stages they also attack maize and sorghum panicles, as well as mainly the ears of maize, generally at the base and in the middle part (Pirkle, et al., 2016).

At establishment, the pest acts as a cutter, when the previous fallow has been kept dirty, with a predominance of grass weeds. When the crop is emerged, it has a preference for the maize head. In this case, the damaged plants recover, but suffer a considerable delay. As a step prior to piercing the head, it damages the leaves with varying intensity depending on the development of its mouthparts (Timoty, et al., 2016).

Quantification of Spodoptera frugiperda damage depends on the level of infestation and the phenological stage of the crop, with the Economic Threshold ranging from 10 to 50 % of infested plants (Timoty, et al., 2016).

The Entomology Section of the National Institute of Agricultural Technology - INTA Pergamino established a scale of damage/degrees:

Grade 1, only gnaw the epidermis of the leaves without perforating them, leaving stains translucent windows known as "ventanitas".

Grade 2, leaf defoliation is moderate and the presence of sawdust or droppings is beginning to be observed (Villamil et al., 2017).

Grade 3, damage to the bud is intense and involves the plant; large larvae and large quantities of excrement are observed. They consume the leaf lamina. producing irregular perforations, and head towards the bud, to feed and protect themselves (Villamil et al., 2017).

The real damage it does to the bud is when it penetrates it. Up to the 4-leaf stage, it eats leaves causing intense damage, but without killing the plant, as the apex or growth meristem is below ground level, but between the 4th and 6th leaf the isoca feeds on the apical primordium and the plant dies. In years of high late infestations it can damage the ear (Villamil et al., 2017).

III. MATERIALS AND METHODS

3.1. Location of the research project

This field research was carried out at the Experimental Farm "La María", which belongs to the State Technical University of Quevedo, located at Km 7 of the Quevedo - El Empalme road, at the following geographical coordinates: 79° 27' west longitude and 01° 06' south latitude, at an altitude of 85 m above sea level.

3.2. Soil pedoclimatic characteristics

The following table shows the edaphoclimatic characteristics of the area where the research was carried out.

Table 2. Edaphoclimatic characteristics of the Experimental Farm "La María". UTEQ - Mocache

Parameters	Average
Average annual rainfall (mm)	2223.80
Average annual temperature (ºC)	26 ºC
Relative humidity (%)	83
Annual heliophany (Hours/light/year)	894.66 hours/light/year
Topography	Irregular relief
Soil pH	6.5

Source: INIAP Agrometeorological Station. National Institute of Meteorology and Hydrology (INAMHI), Pichilingue Tropical Experimental Station, (2021).

3.3. Study factor

3.3.1. Genetic material

The genetic material used were two commercial hybrids (DAS 3385, INIAP 551) and two maize landraces from the INIAP - Pichilingue germplasm bank collection.

3.4. Research design

3.4.1. Research design and statistical analysis

The design and statistical analysis required the use of a randomised complete block design applied to ten treatments with three replications. The Tukey test at 5% probability was used to compare the means of the treatments.

3.5. Treatments under study

As for the mixtures that were sown, these came from four cultivars that have a higher potential in terms of purchase acceptance by producers, which means that two predominant commercial hybrids were used and combined with two landraces that have important characteristics in terms of resistance and tolerance to prevalent health problems. In addition, the monoculture plots were compared with the intraspecific mixture plots, resulting in ten cultivar combinations, as shown in the table below:

Table 3. Combination of hybrids used in the different levels of mixtures and monoculture, according to their characteristics of resistance to phytosanitary problems and their agronomic characteristics.

NO.	Mixtures of maize hybrids and landraces
1	INIAP-551+ maíz criollo Manabí
2	INIAP-551+ maíz criollo Palenque
3	DAS-3385+ maíz criollo Manabí
4	DAS-3385+ maíz criollo Palenque
5	INIAP-551+ maíz criollo Manabí+ maíz criollo Palenque
6	DAS-3385+ maíz criollo Manabí+ maíz criollo Palenque
7	INIAP-551
8	DAS-3385
9	Maíz criollo Manabí
10	Maíz criollo Palenque

Prepared by: Author (2021).

3.6. Delimitation of the experiment

In this section, we proceeded to describe the characteristics of the experiment carried out in the field. As shown below:

Table 4. Characteristics of the experiment to be studied.

Features

Dimensions of each experimental unit	5.0 m * 3.20 m
Area of each experimental unit	16 m2
Distance between maize rows	0.80 m
Distance between plants	0.20 m
Number of rows per plot	4
Number of plants per row	25
Number of plants per experimental unit	100
Number of useful plants per experimental unit	50
Prepared by: Author (2021).	

3.7. Management of the experiment

3.7.1. Soil preparation

Once the area in which the research was carried out was selected, the soil was prepared with two passes of the harrow so that it would be ready for sowing.

3.7.2. Sowing

This phase was carried out manually, using a mirror in which 2 seeds were deposited per stroke, according to a distance between plants of 0.20 m. The seed was also cured with Vitavax 400 of systemic action, before the sowing process so that it would not be damaged by insect or fungus attacks.

3.7.3. Weed control

Two types of weed control were used: The cultural method, which consisted of using management practices that provided the crop with greater management and ease of weed control. On the other hand, chemical control was used, in which a pre-emergent (Glyphosate and Amino 6) was used 15, 30, 45 and 60 days after sowing.

3.7.4. Pest and disease control

No application of agrochemicals for pest and disease control was carried out, due to the nature of the research, so as not to interfere with the variables that were evaluated.

3.7.5. Fertilisation

On the phase of fertilisation was used NPK(Nitrogen (N), Phosphorus (P) and Potassium (K)), at 8, 20, 35 and 60 days after sowing. foliar fertiliser (EVERGREEN) at a dose of 0.5 to 1 litre per hectare, 25, 45 and 65 days after sowing.

3.7.6. Harvest

The harvesting process was carried out manually, when the plants were observed to have reached full physiological maturity after 120 days.

3.8. Variables assessed

3.8.6. Plant height

For the height variable it was measured from ground level to the base of the male panicle. With regard to the samples of the 10 plants taken at random from each useful plot, a ruler was used for the measurement and the unit expressed in metres. This activity was evaluated 60 days after sowing.

3.8.7. Cob insertion height

A total of 10 plants were randomly selected from the useful area of each experimental plot and measured with a tape measure from the base of the stem to the nodule of the ear insertion, and then averaged and expressed as a mean in centimetres.

3.8.8. Weight of 100 seeds

For each treatment and at harvest time, ears were taken at random from each plot in the experimental area and 100 seeds were weighed.

3.8.9. Cob length

The length of the ear was measured from the bottom to the apex, and then averaged and expressed in centimetres. This The activity required taking ten ears at random at harvest time corresponding to each plot in the experimental area.

3.8.10.Grain weight per ear

Ten ears were taken at random from each plot in the experimental area and the kernel weight was recorded and expressed in grams.

3.8.11.Grain-to-cowpea ratio

The kernel-to-crust ratio was determined by taking a sample of ten cobs shelled by hand, dividing the kernel weight by the weight of the crust.

3.8.12.Performance

weight of all the grain obtained in each unit of the experimental plots was recorded, then these values were converted to kg ha^{-1} using a simple rule of three.

3.8.13. Severity levels of foliar diseases (Curvularia lunata, Spiroplasma kunkellii, Puccinia sorghi and Helminthosporium spp.).

Plants were selected to measure the damage caused by foliar diseases, using the arbitrary scale proposed by CIMMYT, and were evaluated at 45 and 65 days of crop age. This scale, showing the values to be considered, is shown below:

Table 5. Arbitrary CIMMYT scale

Scale	Percentage of 0 - 100	Damage
1	0	None
2	0 - 5	Slight
3	5 - 20	Moderate
4	20 - 50	Severo
5	50 - 100	Very severe

Source: CINMYT (2019)

3.8.14.Codling moth severity level by damage/plant

In order to identify, know and evaluate the level of damage caused by (Spodoptera frugiperda), we proceeded to carry out monitoring of the buds of the plants throughout the useful area of the experimental units. This evaluation was carried out by taking twenty plants per treatment, in which the number of healthy and damaged plants was recorded, to the scale in Davis table 6, and they evaluated at twenty and thirty days of age of the crop, taking into account the following aspects:

- Gnawed leaves
- Fresh excreta
- Blade perforations.

Table 6. Scale of damage of Spodoptera frugiperda larvae on leaves.

	Features
1	minimal lesions on the leaves of the head.
2	small holes and circular lesions.
3	small circular lesions and few elongated lesions 1.3 cm.
4	1.3 - 2.5 cm elongated lesions on the leaves of the bud and on unfolded leaves.
5	elongated lesions >2.5 cm and few large small to medium-sized, uniform irregular holes.
6	elongated lesions >2.5 cm with few large holes.
7	many elongated lesions of all sizes and several large orifices.
8	many elongated lesions of all sizes and many large orifices.
9	plant virtually destroyed.

3.9. Statistical analysis of data

The data analysis was performed using an analysis of variance and Tukey's test in order to make a comparison of treatments (hypothesis) and the prediction of a response (dependent variable) from the dependent variables to have a strict compliance, on the basis that, if the data dispersion is high, the data could be adjusted to normality.Given the nature of this experiment, trying to rationalise the information derived from it, the variables were analysed from two different points of view: on the one hand, the situation of the mixtures themselves (the design includes monoculture or a single cultivar/plot, up to the mixture of three genotypes, the number of mixtures will be arranged according to the number of components or genotypes that integrate it, where a statistical analysis was also made according to the number of components).

IV. RESULTS

4.1. Plant height

According to the collection data, the plant height of each of the maize and monoculture mixtures was identified. It was found that the tallest plant with a height of 2.64 metres was the MC Manabí monoculture treatment, followed by DAS + MC Palenque and DAS + MC Manabí + MC Palenque with 2.59 m, and the maize mixture of DAS + MC Manabí obtained 2.58 m, and INIAP + MC Manabí obtained a height of 2.50 m. These plants were the tallest, while the rest were between 2.30 and 2.50 m tall. These plants were taller, while the remaining plants are between 2.30 and 2.45 metres. The difference and impact of different seed characteristics and the way of care and production on the height of the plant is evident.

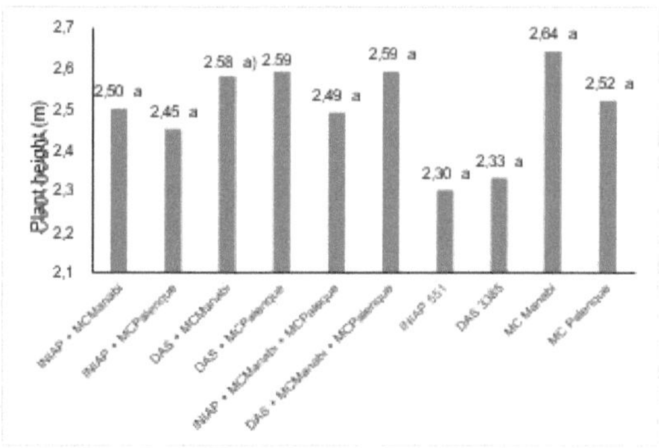

Figure 13. Plant height (m) in different intraspecific mixtures of maize and monocultures.

4.2. Cob insertion height

It can be seen in the results of the experiment that the MC Manabí treatment gave a total height measurement of 139.30 cm of ear insertion, the combination INIAP + MC Palenque with 130.33 cm, the two measurements of INIAP+ MC Manabí obtained 128.84 cm, DAS + MC Palenque measured 126.41 cm. The other treatments, INIAP 551, DAS 3385 and MC Palenque, were below the mean (normal), between 117.38 and 119.18 cm, respectively.

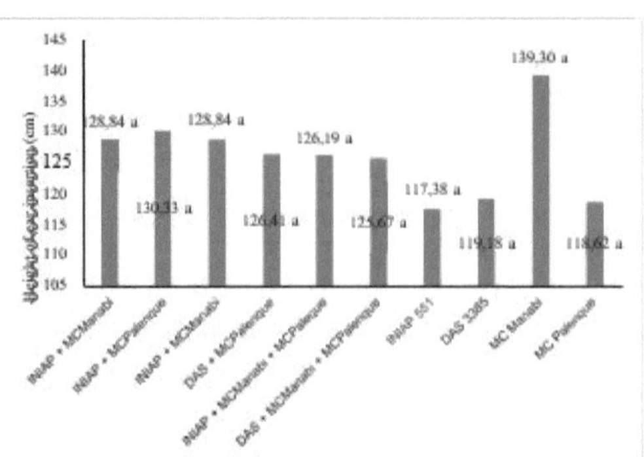

Figure 14. Ear insertion height (cm) in different intraspecific maize mixtures and monocultures.

4.3. Performance components

The results of the table show that the monocultures had a higher average for the characteristics of the ear length component (cm), resulting in the monocultures obtaining a lower average in the yield of grain weight per ear (g), while in the weight of 100 seeds (g) the monocultures had an average percentage and with regard to the grain to ear ratio, the monocultures maintained an average average.

Table 7. Main characteristics of yield components in different intraspecific mixtures of maize and monocultures.

Treatments	Cob length (cm)	Grain weight per ear (g)	Weight of 100 seeds (g)	Ratio grain tusa
INIAP+ MCManabí	14,88 a	96.33 ab	31,67 a	5,07 a
INIAP+ MCPalenque	14,63 a	87,17 b	28,33 a	4,71 a
DAS+ MCManabí	14,76 a	102,00 ab	35,00 a	4,97 a
DAS+ MCPalenque	15,39 a	93.17 ab	33,33 a	4,47 a
INIAP+ MCManabí MCPalenque+	15,05 a	90,33 b	31,67 a	4,32 a
DAS++ MCPalenque	14,93 a	90,83 b	35,00 a	4,57 a
INIAP 551	20,15 a	100.06 ab	33,33 a	4,44 a
DAS 3385	15,10 a	117,33 a	36,67 a	4,91 a
MC Manabí	14,40 a	90,67 b	30,00 a	5,59 a
MC Palenque	15,18 a	96.33 ab	33,33 a	4,61 a
CV (%)	18,69	8,63	12,12	11,73

Note: Tukey's 5% significance test for treatments in the yield assessment separated the averages into six ranges of significance.

4.4. Performance

Figure 15 shows the different treatments with their respective characteristics: measurements (ear length, weight of grain per ear, weight of 100 seeds and grain to seed ratio). In which the high yield of kg ha^{-1} in the monocultures was evidenced, highlighting the Das 3385.

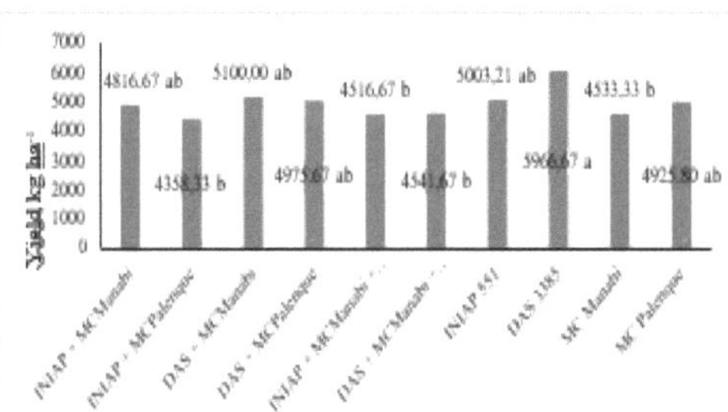

Figure 15. Yield kg ha^{-1} in different intraspecific mixtures of maize and monocultures.

4.5. Incidence of foliar diseases (Curvularia lunata, Spiroplasma kunkellii, Puccinia sorghi and Helminthosporium spp.) in different intraspecific mixtures of maize.

In this section we proceeded to determine the incidence of the foliar diseases C. lunata, S. kunkellii, P. sorghi and Helminthosporium spp. on the different maize intraspecific mixtures, where values and results have been determined and demonstrated as shown below:

a) Occurrence of Curvularia lunata

Figure 16 shows the incidence of Curvularia lunata disease, the level of damage caused by this disease was higher in all monocultures, the same scenario was present at 65 days.

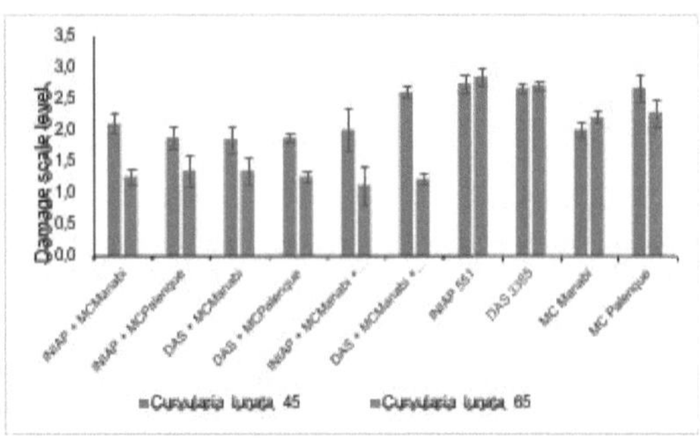

Damage scale level of Curvularia lunata in different intraspecific mixtures of maize and monocultures, with standard error.

According to figure 17, the average damage present in the treatments is follows: in monocultures 45 days the average damage was 3.80 and at 65 days 3.89, in the mixtures of two genotypes the average at 45 days was 2.14 and at 65 days was 2.19, and in the mixture of the three genotypes, the damage at 45 days was 1.83 and at 65 days 1.93. It was concluded that there was damage by this disease in all the components.

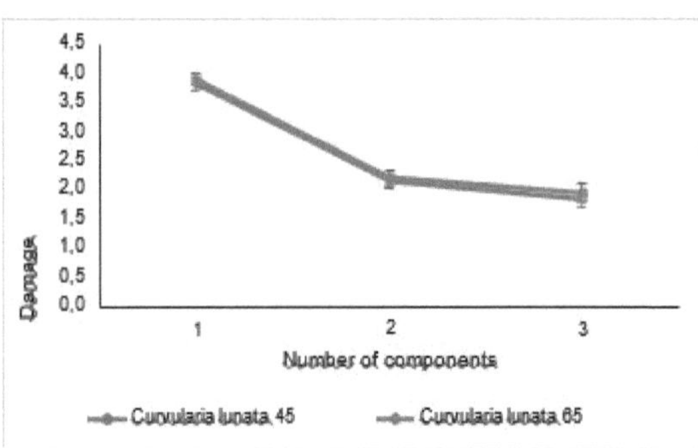

Figure 17. Damage scale level of Curvularia lunata and the number of components used in different intraspecific mixtures of maize and monocultures, with standard error. 1= averages of monocultures; 2= averages of mixtures of two genotypes; 3= averages of mixtures of three genotypes.

b) Incidence of Spiroplasma kunkellii

The level of damage caused by Spiroplasma kunkellii occurred at medium levels in treatments INIAP + MCManabí, INIAP + MCPalenque, DAS + MCManabí, DAS + MCPalenque, INIAP + MCManabí MCPalenque, DAS + MC Manabí + MC Palenque at 45 and 65 days, while MCPalenque, INIAP, MC Manabí, DAS 3385 had higher damage in the same scenario of 45 and 65 days. In this measurement, the damage caused by Spiroplasma kunkellii, commonly known as "red tape", was assessed and the results are as follows:

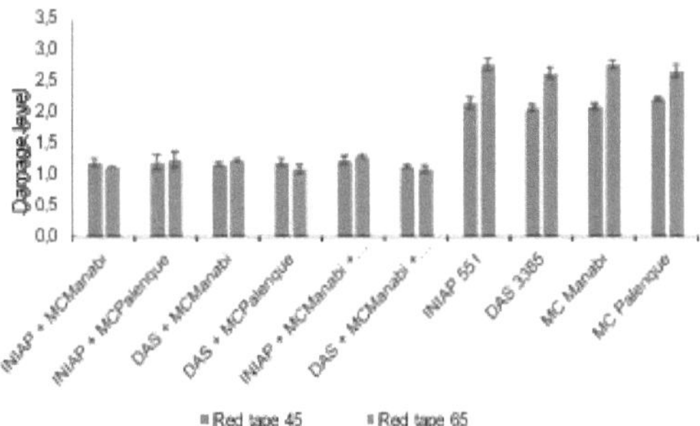

Damage level of Spiroplasma kunkellii in different intraspecific mixtures of maize and monocultures, with standard error.

The averages in figure 19 show that the level of damage decreases according to the number of components in the intraspecific maize mixtures at 45 and 65 days. The following shows the average damage scale level of the foliar disease commonly known as "red ribbon".

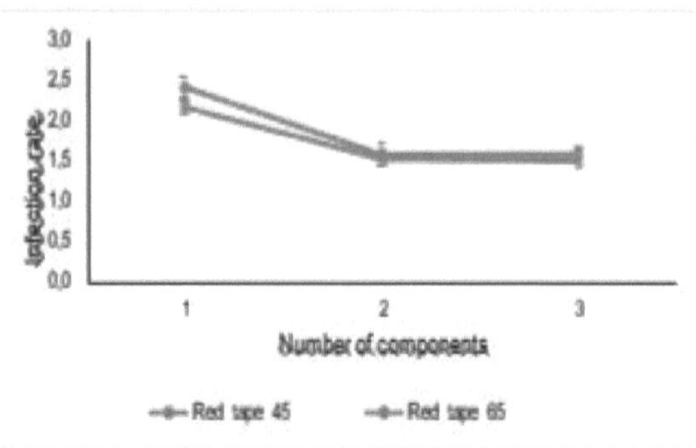

Figure 19. Scale level of damage of Spiroplasma kunkellii and number of components used in different intraspecific mixtures of maize and monocultures, with standard error. 1= averages of monocultures; 2= averages of mixtures of two genotypes; 3= averages of mixtures of three genotypes.

c) **Incidence of Puccinia sorghi.**

The level of damage caused by P. sorghi occurred at lower levels in the treatments INIAP+ MC Manabí, INIAP+ MC Palenque, DAS+ MC Manabí, DAS+ MC Palenque, INIAP+ MC Manabí+ MC Palenque, DAS+ MC Manabí + MC Palenque, whereas, in MC Palenque, INIAP, MC Manabí, DAS 3385 had higher damage, the same scenario occurred at 65 days.

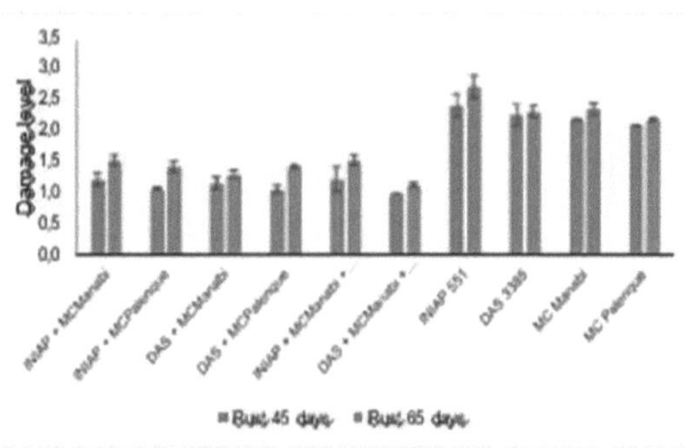

Damage level of Puccinia sorghi in different intraspecific mixtures of maize and monocultures, with standard error.

Figure 21 shows the average rust damage present in the treatments: in the monocultures at 45 days the average damage was 2.23 and at 65 days 2.38, in the mixtures of two genotypes the average at 45 days was 1.57 and at 65 days 1.53, and in the mixture of three genotypes, the damage at 45 days was 1.47 and at 65 days 1.48. An increase was also observed in components 1 and 3, while in component 2 there was a slight reduction at 45 days.

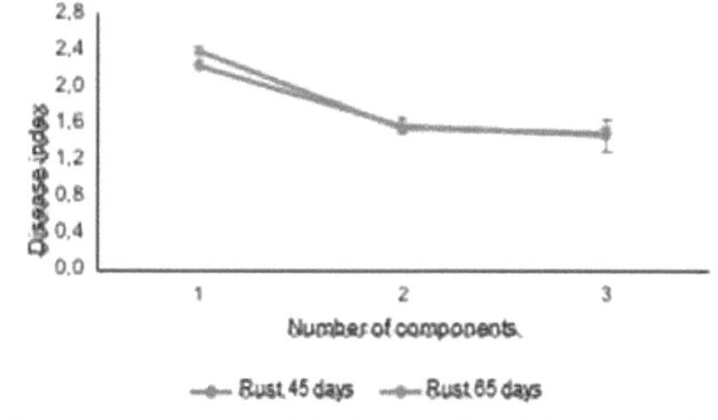

Figure 21. Damage scale level of Puccinia sorghi and the number of components used in different intraspecific mixtures of maize and monocultures, with standard error. 1= averages of monocultures; 2= averages of mixtures of two genotypes; 3= averages of mixtures of three genotypes.

d) **Incidence of Helminthosporium spp.**

This figure clearly shows the level of damage caused by Helminthosporium spp. in two different measurement timelines: 45 days and at 65 days, from which the scale level of damage was severe in all monocultures.

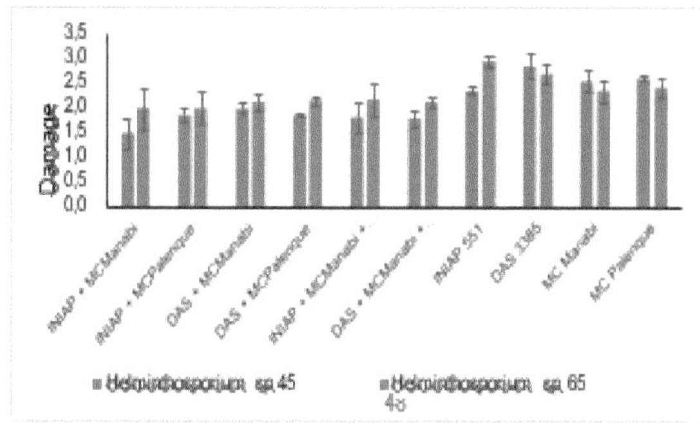

Figure 22. Damage level of Helminthosporium spp. in different intraspecific mixtures of maize and monocultures, with standard error.

The average damage of Helminthosporium spp. in the monocultures at 45 days the average damage was evident and the severe damage increased at 65 days due to the number of components.

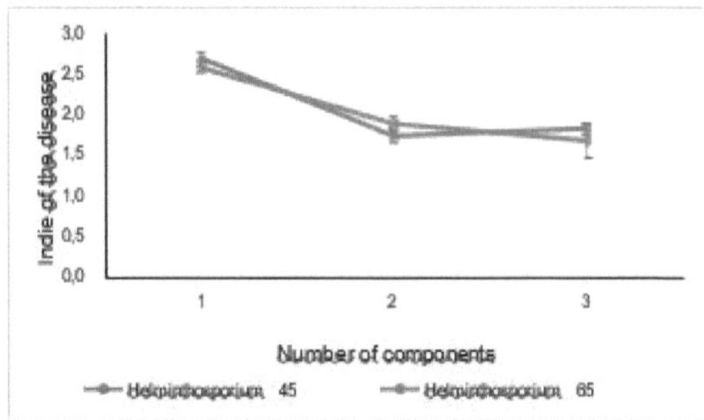

Figure 23. Scale level of damage of Helminthosporium spp and number of components used in different intraspecific mixtures of maize and monocultures, with standard error. 1= averages of monocultures; 2= averages of mixtures of two genotypes; 3= averages of mixtures of three genotypes.

4.6. Damage severity of codling moth (Spodoptera frugiperda) in different intraspecific mixtures of maize.

The severity of S. frugiperda damage was assessed for each treatment on the maize crop. According to the Davis scale, it is categorised as low damage = 1, 2 and 3, medium damage = 4, 5 and 6 and high damage = 7, 8 and 9. The results observed indicate that in the first 20 days all the treatments evaluated were at a low level of damage and at 30 days the damage increased to a medium level in all the monoculture treatments.

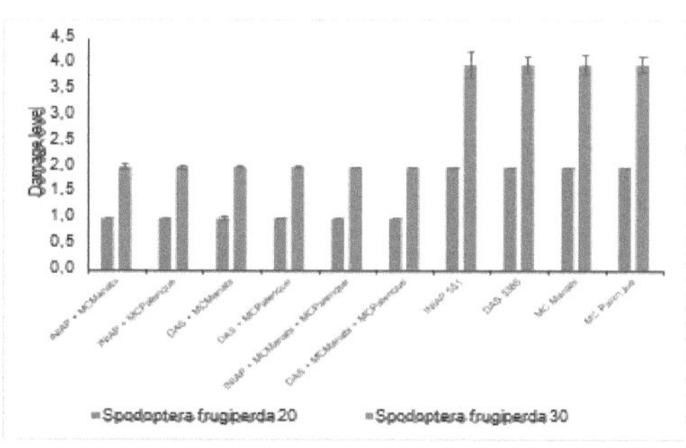

Figure 24. Spodoptera frugiperda damage level in different intraspecific mixtures of maize and monocultures, with standard error. According to Davis scale: low damage 1, 2 and 3; medium damage 4, 5 and 6; medium damage 4, 5 and 6; high damage 4, 5 and 6. high 7, 8 and 9.

It represents the averages of monocultures and obtained a damage level of 2.00, at 20 days and at 30 days it was 4.00. Regarding the mixture of two components it is 1.00 at 20 days and at 30 days the result was 2.00 damage and the mixture of three genotypes obtained at 20 days 1.00 and at 30 days 2.00 damage level in the plant. This shows that the greatest damage was obtained by the monocultures at 30 days, followed by the mixture of the two and three genotypes, which was less than that of the monocultures.

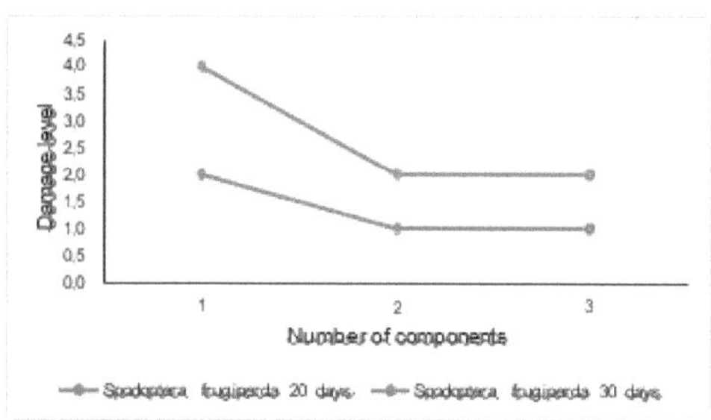

Spodoptera frugiperda damage level and the number of components used in different intraspecific mixtures of maize and monoculture. 1= averages of monocultures; 2= averages of mixtures of two. According to the Davis scale: low damage 1, 2 and 3; medium damage 4, 5 and 6; high damage 7, 8 and 9 genotypes; 3= averages of mixtures of three genotypes. According to Davis scale: low damage 1, 2 and 3; medium damage 4, 5 and 6; high damage 7, 8 and 9.

V. DISCUSSION

Foliar diseases have an impact on yield and numerical components (ear , grain weight per ear, seed weight and grain to ear ratio), as they produce alterations in the systems responsible for production within the plant, such as decreases in the green leaf area index (Alavanja, 2019). The results of the present research corroborate the above mentioned. According to the calculation of yield per kg ha^{-1} to determine the incidence of diseases on the production capacity, it could be observed that in this research diseases negatively affect yields, this situation coincides with Quintana (2019) who states that "foliar diseases are one of the main biotic factors that limit the expression of crop yields". The level of incidence of foliar diseases in different intraspecific mixtures, which were INIAP + MCManabí, INIAP + MCPalenque, DAS + MCManabí, DAS + MCPalenque, INIAP + MCManabí + MCPalenque, INIAP + MCManabí + MCPalenque, INIAP + MCManabí + MCPalenque, DAS + MCManabí + MCPalenque, as well as monocultures INIAP 551, DAS 3385, DAS 3385, MC Manabí, MC Palenque. With regard to Curvularia lunata, Spiroplasma kunkellii, Puccinia sorghi and Helminthosporium spp. decrease in the level of damage was evident due to the number of their components.It has been possible to observe and analyse that foliar diseases negatively affect the maize crop, and that the growing increase is due to the deficiency in the management of these diseases, which is why it is necessary to make a proposal as an alternative, which required data from previous sections in which it was demonstrated both the incidence of pests and diseases, as well as the damage caused by the pest budworm to the maize crop that was subjected to this experiment. research with the aim of making a positive contribution to the treatment and production results of the plant. According to the results regarding the mixtures of various genotypes (DAS 3385, Maíz criollo Manabí and Maíz criollo Palenque) it is identified that when mixing the genotypes, they are influenced by the crossing of resistance genes from criollo materials to hybrids and vice versa, positively affecting the agronomic development and resistance to pests and diseases, this gives us the guideline to recommend this type of alternative maize crop production where the applications of phytosanitary products will be minimised. It was possible to identify the severity of damage of the budworm in the 10 treatments applied in this research, in which it was determined in two measures of time at a general level: 20 and 30 days to make their respective measurement in which it was shown that the level of damage caused at 20 days was low, but as the time increased

(30 days) the level increased to medium level. In this regard, Louette and Smale (2016) state that infestations of this pest cause severe damage to crops, which directly leads to yield reductions of more than 30%. Therefore, great care must be taken in crop management. Regarding the average damage of S. frugiperda present in the treatments and according to the components used in the maize crop, it was obtained that the greatest damage was obtained by the monocultures, as the mixture of the two genotypes, as well as those of the three genotypes was less damage caused by diseases and pests; which shows as a consequence that the intraspecific mixture of seeds can positively affect the severity of damage caused by this common pest in the maize crop. This coincides with Yánez et al., (2018) who state that intraspecific mixtures are established for breeding, feeding and protection against pests and diseases and this makes them more resistant and does not considerably affect the crop yield.All the information in this work is directed to the importance of giving a correct management and care to the different crops, especially maize, which is the most important one. which is constantly attacked by pests and diseases that affect production. The reduction of damage due to the mix of genotypes and monocultures used, allows the farmer to take action in order to avoid situations that directly affect the crop, as well as his economy due to the reduction in yields. With regard to the application of different intraspecific mixtures in maize and the results on the resistance they showed to pest and disease attacks, the importance of applying them to maize crops is noted. Barg and Armand (2017) also agree, stating that the use of intraspecific mixtures and different maize genotypes partially prevents pests from damaging the crop and therefore avoids affecting production yields, but it is also vital because the use of agroecological methods can solve problems related to erosion and better conserve the soil and increase the stability of agricultural systems. Intraspecific mixtures applied to maize generate positive effects on the crop's entomofauna, which in turn allows it to be properly managed and not only protects production but also the soil, water regulation, biodiversity conservation and, of course, the reduction of costs in many labour tasks generally carried out by farmers (Sánchez and Ruiz 2016). Based on this analysis, it can be concluded that the application of intraspecific mixtures for the control and care of pests and diseases present in maize crops would have a positive effect on both production and the preservation of the ecosystem.

VI. CONCLUSIONS

- The use of different maize intraspecific mixtures of the different treatments applied in the experiment allows a reduction in the incidence of the foliar diseases Curvularia lunata, Spiroplasma kunkellii, Puccinia sorghi and Helminthosporium spp.
- The intraspecific mixture of DAS 3385, Manabi creole maize and Palenque creole maize reduces the attack of codling moth (Spodoptera frugiperda).
- The implementation of different intraspecific mixtures using several maize genotypes allows high yields kg ha^{-1}.

VII.RECOMMENDATIONS

- Carry out a more in-depth genetic analysis, using molecular markers to identify and select the best maize genotypes with greater precision in order to ensure more effective maize production.
- Use the material collected in this research to develop breeding programmes for genotypes for future use and to obtain more detailed information for the benefit of the producer/farmer.
- It is recommended to use the mixture Das 3385, Manabi Creole maize and Palenque Creole maize as an alternative to the attack of foliar diseases. C. lunata, S. kunkellii, P. sorghi and Helminthosporium spp, and the codling moth Spodoptera frugiperda.

VIII. BIBLIOGRAPHICAL REFERENCES

Aguirre, A., Bellon, M., and Smale, M. (2019). A regional analysis of maize biological diversity in Southeastern Guanajuato, Mexico. CIMMYT Economics Working, 89.

Allen, T. (2016). Virtual Herbarium and Maize Phytopathology: Diseases and Pests of Maize, Retrieved from https://herbariofitopatologia.agro.uba.ar/?page_id=162

Alavanja, M. (2019). Pesticides use and exposure extensive worldwide. Rev.

Environ Health, 29(4), 303- 309.

Arévalo, P., and Mejía, I. (2018). Management of the budworm. Retrieved from www.soccolhort.com/revista/pdf/vol1/art9

Arregui, M., and Puricelli, E. (2018). Mechanisms of Action of Pesticides. Dow Agrosciences.

Bernardino, H. U., Torres, H., Sánchez, G., Reyes, L., and Zapién, A. (2019). Pesticide use in maize cultivation in rural areas of Oaxaca State, Mexico. Rev. environmental health, 19(1), 23-31.

Brush, B. (. -1. (2018). Genetic diversity and conservation in traditional farming systems. Journal of Ethnobiology 6, 153-165.

Caballero, P., Murillo, R., Muñoz, D., and Williams, T. (2019). Spodoptera frugiperda (Lepidoptera: Noctuidae) nucleopolyhedrovirus as a biopesticide. Revista Colombiana de Entomología, 2(35), 105-115.

Caicedo, M. (2017). Current situation of maize cultivation in Ecuador. Lima, Peru: INIEA.

CIMMYT. (2019). Maize seed industries, revisited: Emerging roles of the public and private sectors. Mexico: CIMMYT.

Corra, L. (2019). Training tools for the responsible handling of pesticides and their containers: health effects and exposure prevention. Buenos Aires: Pan American Health Organization - PAHO.

Devine, G., Eza, E., Ogusuky, E., and Furlong, M. (2018). Insecticide use: context and ecological consequences. Rev. Peru Med Exp Public Health, 2(1), 74-100.

Duke, S. O. (2016). Herbicide-Resistant Crops: Agricultural, Environmental, Economic, Regulatory and Technical Aspects. Boca Raton, Florida: CRC Press, Lewis Publ.

Gómez, G., and Minelli, M. (2017). La producción de semillas: Texto básico para el desarrollo del curso de Producción de semillas en la Universidad de Nicaragua. Managua, Nicaragua: Instituto Superior de Ciencias Agropecuarias, Escuela de Producción Vegetal.

González, B. (2016). The Green Revolution in Mexico. Revista Agraria, 40-68.

Goodman, G., and Rawling, J. O. (2018). Appropriate characters for racial classification in maize. Botanical Economics, Vol. 47, 44- 59.

Guerrero, A., J. Florián, M., and Florián, J. (2017). Fertilizer and pesticide use in the district of Poroto, Trujillo-La Libertad. Sciendo, 91-102.

Kundu, S. R., Ved Prakash, H., Gupta, H. P., and Ladha, J. (2017). Long-term yield trend and sustainability of rainfed soybean -wheat system through farmyard manure application in a sandy loam soil of the Indian Himalayas. Biology & Fertility of Soils, 271-280.

Louette, D., and Smale, M. (2016). Genetic diversity and maize seed management in a traditional Mexican community: Implications for in situ conservation in maize. Mexico: NGR.

Maffei, J., and Bueno, L. (2018). Soils and Fertilization. Olive growing in Mendoza. Raigambre de una actividad que se renueva. Buenos Aires: Ed. Pedro Marzano Foundation.

Maya, N. (2016). Evaluation of seven maize (Zea mays L.) genotypes at four locations in Nicaragua. Postgraduate work. Managua, Nicaragua: Universidad Nacional Agraria.

Naranjo, A. (2017). Blurring the agrarian model. The case of maize. Mexico.

Navarrete, C. L. (2017). Effect of cultural management of an intraspecific mixture system of musaceae on the incidence and severity of the main phytosanitary problems (Bachelor's thesis, Quevedo-UTEQ). Quevedo: Quevedo State Technical University.

Ortega, R. (2017). Maize as a crop II. Maize diversity in Mexico.

Mexico City: Dirección General de Culturas populares e Indígenas.
PAN International. (2016). Consult Manual. Pesticide Action Network International. California, USA.

Pesticide Action Network (2018). List of Highly Hazardous Pesticides.

PAN International.
Pirkle, J. L., Sampson, E. J., Needham, L. L., Patterson, D. G., and Ashley, D. L. (2016). Using biological monitoring to assess human exposure to priority toxicants. Environ Health Perspect, 45-48.

Poehlman, J., and Sleper, D. (2015). Crop Genetic Improvement.

Mexico: Limusa.
Quintana, W. (2019). Maize genetic diversity for biosafety purposes.

Portoviejo: Universidad Técnica de Manabí.
Ramírez, J., and Lacasaña, M. (2018). Pesticides: classification, use, toxicology and exposure measurement. Arch. Prevención Riesgos Laborales. Barcelona. Spain. 4 (2): 67-75. Revista Agroecológica, 4(2), 67-75.

RAP-AL (2017, September 23). What are pesticides. Retrieved from http://www.rap-al.org/index.php?seccion=4&f=plaguicidas.php

Rendón, B. A., and Aragón, M. (2017). Maize diversity in the Sierra Sur de Oaxaca, Mexico: traditional knowledge and management. Polibotánica, 151-74.

Rimache, M. (2018). Cultivo del Maíz. Venezuela: Espasandes.
Romero, Y. (2019, February). Biological control of maize budworm with baculovirus. Retrieved from htpps://www.researchgate.net/publication/32926352

Salazar, N., and Aldana, M. (2018). Glyphosate herbicide: uses, toxicity and regulation.

Biotechnology, 13(2), 8-23.
Sánchez, J. J., and Ruiz, C. (2016). Gene flow between criollo maize, improved maize and teocintle: implications for transgenic maize. Mexico: Trillas.

Tielemans, E., Van Kooij, E., Velde, B. A., & Heederik, D. (2017). Pesticide exposure and decreased in vitro fertilisation rates. The Lancet, 484-485.

Timoty, D., Hathenway, W., Grant, U., Torregroza, M., Sarria, D., and Vela, D. (2016). Maize breeds of Ecuador. Technical Bulletin No. 12. Colombia: ICA.

Villamil, E., Bovi, G., and Nassetta, M. (2017). Current situation of pesticide contamination in Argentina. Rev. Int. Contam. Amb., 19(1), 25-43.

Yánez, C., Zambrano, J., and Caicedo, M. (2018). Maize Production Guide for

small . Quito, Ecuador: INIAP.

Yanggen, D., Crissman, C., and Espinosa, P. (2019). Pesticides, Impacts on production, health and environment in Carchi, Ecuador. Carchi, Ecuador: Universidad Técnica Particular de Loja.

TABLE OF CONTENTS

Printed by Books on Demand GmbH, Norderstedt / Germany